Semirames Silva
Eliezer Siqueira
Danielle Nascimento

Moringa oleifera: A viable alternative for water treatment

Semirames Silva
Eliezer Siqueira
Danielle Nascimento

Moringa oleifera: A viable alternative for water treatment

ScienciaScripts

Cover image: www.ingimage.com

This book is a translation from the original published under ISBN 978-3-330-99702-8.

Publisher:
Sciencia Scripts
is a trademark of
Dodo Books Indian Ocean Ltd. and OmniScriptum S.R.L publishing group

120 High Road, East Finchley, London, N2 9ED, United Kingdom
Str. Armeneasca 28/1, office 1, Chisinau MD-2012, Republic of Moldova, Europe
Printed at: see last page
ISBN: 978-620-8-19732-2

Moringa olelfera: A viable alternative for water treatment

Authors: Semirames do Nascimento Silva; Danielle Maria do Nascimento; Eliezer da Cunha Siqueira

1. INTRODUCTION

Water is directly responsible for maintaining life and the well-being of the human species, as well as being the most important natural resource for the use of other resources, whether plant, animal or mineral. Without water there would be no conditions for the development of life on Earth. It is the basic liquid for meeting man's basic needs and is present in all living beings. Access to good quality water in adequate quantity is directly linked to the health of the population, helping to reduce the occurrence of various diseases.

According to the World Health Organization (WHO), all people at all stages of development and socioeconomic conditions have the right to safe drinking water in sufficient quantity to meet all basic needs (PANAMERICAN HEALTH ORGANIZATION, 2014).

With the quality and quantity of fresh drinking water available to supply the world's population compromised, alternative treatments are needed to ensure the quality of the treated water.

The service of supplying water through a general network is characterized by taking raw water from nature, adapting its quality, transporting it and supplying it to the population through a general distribution network. Alternative ways of supplying the population must also be considered, such as water from fountains, private wells, water tankers, cisterns and others (IBGE, 2010).

In Brazilian urban areas, 90.4% of households have access to a water distribution network, while in rural areas only 16.7% of households have this system (BRASIL, 2006). It is estimated that around 12 million Brazilians do not have piped or treated water for residential supply. In the northeast region, 21.8 and 24.3% of the municipalities in the states of Maranhao and Piaui do not treat the water distributed to the population. The regional approach shows that, in 2008, of the 33 municipalities without a general water supply network in at least one district, 21 (63.3%) were located in the Northeast region.

Northeast, especially the states of Paraiba (11 municipalities) and Piaui (5); and seven (21.2%) in the North, especially the state of Rondonia (4 municipalities) (IBGE, 2010).

According to Libanio (2008), water treatment systems for drinking water are made up of a flocculation tank (using aluminum sulphate in the coagulation/flocculation process), a decantation tank (removal of flocs), a filtration system (removal of particulate matter) and a disinfection system (using chlorine, ozone or ultraviolet radiation to inactivate microorganisms).

There are more and more environmental and human health contraindications to the use of chemical compounds in the treatment of wastewater or in the production of drinking water. Aluminum sulphate stands out as the most widely used chemical coagulant in Brazil, due to its good efficiency and low cost. However, as aluminium is not biodegradable, high concentrations of this compound can cause problems for human health, including accelerating the degenerative process of Alzheimer's disease (CLAYTON, 1989). In this context, an environmentally friendly and affordable coagulant is a viable alternative for water treatment. Thus, numerous plants are being used as natural coagulants and/or flocculants, such as *Moringa oleifera* Lam. Moringa *(Moringa oleifera* Lam.) is a well-known plant in northeastern Brazil, especially in the semi-arid region.

Water clarification by sedimentation using natural coagulants, alternatives to metal salts, stand out for being a process that does not require the use of other products

In addition to being one of the simplest water treatment processes, it does not require specialized labor and is affordable for small communities, especially in rural areas.

The importance of alternative clarification techniques, such as the use of natural coagulants that improve water quality, should be emphasized. This alternative is an instrument of unquestionable importance, since it can also guide support and proposals aimed at the efficient use of water resources and improving the ways in which water is used in all its aspects, both in terms of quantity and quality.

2. OBJECTIVES

2.1. General Objective

Obtaining a natural coagulant from moringa seed powder and using it to treat water in rural cisterns.

2.2. Specific Objectives

- Evaluate the chemical composition of the seeds: total protein (%), lipids (%), ash (%) and moisture (%);
- Evaluate the efficiency of using the coagulant at a concentration of 2.5 ml for 30 minutes, assessing: electrical conductivity, pH, probable residue, $CaCO_3$, NaCl and total alkalinity;
- Evaluate the microbiological quality of the water, such as fecal *coliforms* and *Escherichia coli.*

3. LITERATURE REVIEW

3.1. Conventional water treatment

The main operations carried out in conventional water treatment are: coagulation, flocculation, sedimentation, filtration and disinfection. Coagulation is a crucial process in the removal of dissolved and suspended particles in water, consisting of the destabilization of colloidal particles that can be removed in the following stages (LILLIEHOOK, 2005).

The conventional water process employs sedimentation using coagulants and comprises the following unit operations: coagulation, flocculation, decantation and filtration to clarify the water, followed by pH correction, disinfection and fluoridation (BOTERO, 2009).

This is a chemical treatment that allows for the subsequent deposition of suspended substances and colloids present in the water, using a compound such as aluminum sulfate or ferrous sulfate. This process is carried out as a rapid mixture, in which the chemical compounds are added when the water is turbulent due to the presence of structures such as flumes or spillways (FUNASA, 2003). According to Lilliehook (2005), coagulation and flocculation can remove 99% of bacteria and viruses.

An important stage in water treatment is disinfection, which aims to destroy or inactivate the vast majority of pathogenic organisms capable of producing disease, or other undesirable organisms. The most common chemical agent used in the disinfection of water supplies and wastewater is chlorine. Despite the benefits of disinfection, chlorine and other oxidizing compounds can react with natural organic matter (NOM) from surface water sources, forming disinfection by-products that can be harmful to human health, such as trihalomethanes (THM), haloacetic acids (HAA) and haloacetonitriles (HAN), among others (HONG, 2007).

In conventional water treatment plants, the water is agitated followed by the addition of coagulants to reduce the repulsive force between the particles, increasing their collision and the formation of flocs (BORBA, 2001). The coagulant undoubtedly plays a key role in this process. According to Gassen et al. (1990) coagulation and/or flocculation is a common process in water treatment, which stabilizes suspended and dissolved materials, followed by aggregation into large flocs that are not easily separated from the water by subsequent sedimentation and filtration processes. Finally, disinfection eliminates pathogenic microorganisms that are still present in the water.

Many coagulating agents are used in water treatment processes, such as inorganic coagulants like aluminum and iron salts, synthetic and natural organic polymers. Due to their high coagulating capacity and low cost, aluminum and iron salts are the most widely used agents in water treatment (CARDOSO, 2007).

According to Borba (2001), the most common changes currently used in conventional water treatment are: water treatment with direct filtration, which consists of adding a small amount of coagulant to the raw water to destabilize the impurities and then filtering it.

In this modality, the decantation phase is eliminated and takes place in a simplified stage, usually consisting of a coagulation, flocculation and filtration system, with an ascending or descending filter, followed by disinfection or chlorination, the water treatment is done in a floc-decanter, where the coagulation, flocculation and decantation stages are compacted into a single stage called floc-decantation.

These conventional water treatment processes can be simulated in the laboratory by means of treatability tests in equipment such as the Jar test, which is a set of static reactors that provide different velocity gradients. In this way, different coagulants can be tested on the water to be treated and flocculation and sedimentation can also be simulated, with the water analyzed before and after these operations (DI BERNARBO; DANTAS, 2005).

The coagulation of turbid water is defined as the water treatment process that aims to form electrically unbalanced particles. The occurrence of the phenomena of coagulation and/or chemical flocculation plus decantation or sedimentation leads to clarification, which is defined as the process of removing color and turbidity from natural waters.

The clarification of water can be defined as the removal of impurities that are dispersed in it, giving it the characteristics of raw water such as apparent color and turbidity that prevents light from passing through it, making it opaque and non-translucent.

According to Di Bernardo (1993), the conventional coagulants most commonly used in water treatment are: aluminum sulphate, ferrous chloride, ferric chloride, among others. Due to the high electropositivity of the chemical elements that make them up, when they are dissolved in water, they generally form gelatinous compounds with positive charges, known as cationic coagulants.

In conventional water treatment, aluminum-based coagulants, such as Poly Aluminum Chloride (PAC), are widely used due to their proven efficiency and low cost. However, research has pointed out some disadvantages, such as health problems, such as Alzheimer's disease, caused by residual aluminum in treated water, the production of a large volume of sludge, consumption of the alkalinity of the medium, leading to additional costs with chemical products used to correct the pH of the treated water (RONDEAU et al., 2000; KATAYON et al., 2006). Therefore, it is necessary to look for alternative and more environmentally compatible coagulants.

3.2. Natural coagulants used in water treatment

Brazil has one of the largest reserves of water resources in the world. However, many of the water sources used are increasingly polluted and deteriorating due to a lack of control or investment in the collection, treatment and final disposal of sewage and industrial effluents (CARVALHO, 2008). There are more and more environmental and human health contraindications to the use of chemical compounds in the treatment of wastewater or in the production of drinking water.

Currently, the use of natural coagulants such as *Moringa oleifera* Lam. has been studied as an option to conventional water treatment. According to Lo Monaco et al. (2010), the use of these coagulants can alleviate the problems associated with the consumption of non-potable water and the discharge of untreated wastewater into receiving water bodies.

Natural coagulants of plant origin were used for water treatment before the advent of chemical salts and their use was progressively diminished (NDABGENGESERE; NARASIAH, 1998). There has been a resurgence of interest in coagulants because they are biodegradable and safe for human health. In studies carried out by Ozacar and Sengil (2002), it was found that tannins are high molecular weight polycyclic aromatic compounds, found frequently in plants, and are an efficient primary coagulant.

Natural coagulants of organic origin, known as polyelectrolytes, are compounds made up of large molecular chains with positively or negatively charged sites (BORBA, 2001). Zhang et al. (2002) studied the coagulant properties of the cactus in water treatment and several studies have been carried out to evaluate the activity of coagulant compounds present in moringa seeds (GHEBREMICHAEL et al., 2005).

The importance of alternative clarification techniques, such as the use of natural coagulants that improve water quality, should be emphasized. This alternative is an instrument of unquestionable importance, since it can also guide support and proposals aimed at the efficient use of water resources and improving the ways in which water is used, both in terms of quantity and quality.

According to Jahn (1989), the coagulant product present in moringa seeds was isolated and six polypeptides were identified, which are made up of various amino acid units. The active fraction of this coagulant is due to the presence of a high molecular weight cationic protein, which destabilizes the particles contained in the water and coagulates the colloids (NDABIGENGESERE et al., 1995). According to Gallao et al. (2006), protein is the compound found in greatest quantity in moringa seeds, at around 40%.

When the seed powder is dissolved in water, it acquires a positive charge that attracts negatively charged particles such as clays and silts, forming dense flocs that sediment. The moringa seed-based coagulant, being of natural origin, has a significant advantage over the chemical coagulant, aluminum sulphate, especially for small communities since it can be prepared on site (AMAGLOH; BENAG, 2009).

Interest in the study of natural coagulants to clarify water is not a new idea. The study and research of water treatment with moringa received a major boost at the end of the 20th century (JAHN, 1988). The main interest of researchers studying coagulation and moringa revolves around pH variations, the use of different parts of the plant, different methods and solvents to obtain the coagulant solution.

According to Ndabigengesere and Narasiah (1996), moringa seeds are a viable alternative coagulant agent to replace aluminum salts, which are used in water treatment all over the world. Compared to aluminum, moringa seeds did not significantly alter the pH and alkalinity of the water after treatment and did not cause corrosion problems.

Natural coagulants can be used to treat water for human consumption, due to the fact that they do not alter the alkalinity of the effluent and eliminate micro-organisms that can be harmful to human health, such as the *E. coli* bacteria (BINA et al., 2009).

Bhole (1995) studied the coagulant substances of three plants native to India: *Myrobalan (Terminilia chebula), Beleric myrobalan* and *Emplica myrobalan (Morinda citrifolia)* which, when used as a primary coagulant, proved to be efficient for a wide range of turbidities (25 to 1600 NTU).

The aqueous extracts of dried moringa seeds proved to be compatible with aluminum salts and their use was recommended as a coagulating agent in water treatment in developing countries (JAHN, 1988). Similar studies have also been carried out with other plant materials, such as tamarind seeds (BHOLE, 1995).

In 1974, Reinhold Kickuth devised the first plant-based treatment stations that were successful. Experiments were carried out with dogwood *(Phragmites autralis).* This grass from the banks of rivers and ponds has two very important advantages for use as the main plant in polluted water treatment plants: rapid growth and an efficient system for transporting oxygen from the leaves to the roots. It is now known that the collaboration between plants and soil mixtures creates a type of reactor where the biological and chemical load of the water used is treated until the polluting components are mineralized (LORGE et al., 2011).

According to Carrijo (2002) there is already work on the use of coagulants based on polysaccharides, proteins and mainly starch, in particular cassava flour, arrowroot and potato starch.

As with all coagulants, the effectiveness of the seeds can vary from one water to another. Studies have been carried out to determine potential risks related to the use of moringa seeds in water treatment. To date, no evidence has been found that the seeds can cause side effects in humans, especially at the low doses required for water treatment.

Natural coagulants have shown advantages over chemical coagulants, specifically in terms of biodegradability, low toxicity and low sludge production rates (MORAES, 2004). According to Santana (2009), the use of natural coagulants of plant origin to clarify cloudy and colored water is of great ecological and environmental importance.

To this end, tannins act in colloidal particle systems, neutralizing charges and forming bridges between these particles, and this process is responsible for the formation of flocs and consequent sedimentation (GRAHAM et al., 2008). Tannin does not alter the pH of the treated water as it does not consume alkalinity from the medium, while it is effective in a pH range of 4.5 - 8.0 (BARRADAS, 2004).

In order to minimize problems in water and sewage treatment plants in small and large cities, and In large projects to clean up rivers and lagoons in urban areas, it is being proposed to use synthetic inorganic or organic polymers instead, or aluminium or iron salts, the use of organic cationic polymers prepared from natural products, such as tannin extracted from *Acacia mearnsii* de Wildemann, or black acacia, a plant of Australian origin, is being proposed to provide sustainable support for water treatment in Brazil. Black acacia is grown in Brazil only in the state of Rio Grande do Sul (MANGRICH, et al., 2013).

The use of natural coagulants combined with other tertiary wastewater treatment processes, such as adsorption, which removes compounds of organic origin and metals, and the ultrafiltration process used to remove organic matter and microorganisms, have been shown to be alternative techniques that also allow wastewater to be reused (FUGLIE, 2001).

Water and sewage treatment technologies must not jeopardize the quality of life of future generations, must be suited to the local culture so as not to drastically interfere with social traditions and must not be dependent on external mechanisms for the transfer of equipment and other inputs.

The use of biodegradable organic coagulants is therefore a technical alternative to conventional coagulants, with benefits for public health as well as environmental preservation (BONGIOVANI et al., 2010).

3.3. *Moringa oleifera* Lam

Moringa oleifera Lamarck is a perennial species of the Moringaceae family, native to northeastern India, consisting of only one genus Moringa and 14 species widely distributed in India, Egypt, the Philippines, Pakistan and other countries (KARADI et al., 2006). It is a tree with a thick, tall stem of up to 10 meters. Its leaves are long-petiolate, bipinnate, with obovate leaflets up to 3 cm long (SILVA; MATOS, 2008). The seeds are dark brown with three wings and a paper-like appearance. The main root is thick, it flowers and produces fruit and seeds all year round (RARNACHANDRAN et al., 1980).

Moringa can be easily propagated by seeds or cuttings. The seeds can be sown directly in the final location or in seedbeds. No prior treatment is required. The plant requires little cultivation.

In Brazil, in some regions this species is already widespread and used as an alternative solution for clarifying the water supply of isolated rural homes, including in the state of Paraiba (BORBA, 2001). It grows in regions ranging from the dry and humid subtropics to the dry tropics and humid forests. It is drought tolerant, flowering and producing fruit during this period (DUKE, 1978).

According to Dalla Rosa (1993), moringa adapts to a wide range of soils, but grows best in well-drained black soil or clayey black soil, preferring neutral to slightly acidic soil. The plant produces fewer leaves when it is continuously under water stress. In places where rainfall is below 300 mm, it requires a relatively high phreatic soil to produce (DALLA ROSA, 1993).

Recent studies show that these by-products can also be used as filter media in produced water treatment, such as oil extraction, for example. Their seeds contain a water-soluble substance with coagulating properties (MUYIBI; EVISON, 1995).

The seeds can also be used to treat water by flocculation and sedimentation. Moringa is characterized by being very drought tolerant, being cultivated in arid and semi-arid regions, which have annual rainfall of less than 300 mm (PEREIRA, 2010), making it a very economically viable crop. They are traditionally used to treat water for domestic use in Sudan and Indonesia (PARENTE, 2003).

According to Pereira et al. (2010), *M. oleifera* seeds produce an excellent oil that can be used for food, soap making, cosmetics and the production of high-quality biodiesel. According to Pereira (2010), the oxidation stability of moringa oil is exceptional, reaching 93 hours.

Its flowers are highly sought after by bees (GERDES, 1996). The oil contained in the seeds has a high food and industrial value. It is clear, sweet, odorless and resistant to rancidity. Moringa seeds contain 38% of their weight in oil, which is made up of oleic (76%), linoleic (3.1%), palmitic (8.3%) and stearic (8.0%) acids (DAHOT, 1998).

According to Rashid et al. (2011) the average oil content of moringa seeds is higher than that of conventional oil crops such as cotton (15 to 24%), soy (17 to 21%), safflower (25 to 40%) and mustard (24 to 40%), grown in countries such as the United States, Brazil, China and other Asian and European countries.

Moringa has great potential due to its many food, agricultural, medicinal and industrial uses (LORENZI; MATOS, 2002). Extracts of the leaves have hypotensive and anti-ulcer hypocholesterolemic activity and the seeds have antifungal and antibacterial activity (VASCONCELOS et al., 2009). According to Rashid et al. (2011) the fruits, seeds, leaves and flowers have excellent nutritional value and are consumed in both human and animal food.

The cosmetics industry also uses moringa seed because of the large amount of oil present in its seed. The roots are considered stimulants and diuretics. Externally, it is used for inflammation. The tea from the flowers is widely used in Haiti to cure colds (OKUDA et al., 1999).

Moringa has seven times more vitamin C than oranges, four times more calcium than milk, four times more vitamin A than carrots, three times more potassium than bananas, twice as much protein as milk (about 27% protein, equivalent to beef), more iron than spinach, vitamins present: A, B (thiamine, riboflavin, niacin), C, E, and beta carotene. And minerals such as phosphorus, iron, selenium and zinc (SILVA, 1999).

Moringa seeds can be used to fight the larvae of the Aedes aegypti mosquito, which transmits dengue fever and yellow fever. According to a study carried out at the Federal University of Pernambuco, one of the proteins contained in the seeds, lectin, prevents the larvae from digesting, causing them to die from malnutrition (RANGEL, 2004). The diverse uses and potential of moringa have attracted the attention of researchers, extension workers, development agencies and producers in all major regions of the world.

4. METHODOLOGY

4.1. Determining the chemical composition of the seeds of *Moringa oleifera Lam.*

The chemical composition of the moringa seeds included protein, lipids, ash and moisture. The analyses of the chemical composition of the seeds were carried out in the Physico-Chemical Analysis Laboratory of the Agroindustry and Food Technology Department of the Federal Institute of Education, Science and Technology of Paraiba - IFPB Campus Sousa.

Table 1. Results of the chemical composition of moringa seeds. Sousa-PB, 2016.

Analysis	**Result (%)**
Proteins	34,22
Lipids	39,84
Ashes	3,35
Humidity	5,02

4.2. Obtaining moringa seed powder and preparing the coagulant

To obtain the seed powder, the seed was macerated in a glass container until it became a fine paste. The moringa coagulant was obtained in an aqueous medium. 150 mL of distilled water and 1.5 g of moringa seed powder were added to a beaker. The solution was then homogenized using a mixer at a speed of 1000 rpm for 30 minutes.

4.3. Sample preparation and treatments

The water used in the work came from cisterns built in the Santo Antonio settlement, in the municipality of Cajazeiras - Paraiba. The source of the water in the cisterns during the period in which the work was carried out (June/December 2014) for C1, C2 and C3 was rainwater and for C4 and C5 it was water supplied by water tankers. The water used in the experiment was analyzed before and after treatment in the Microbiology Laboratory of the IFPB Campus Sousa to check for: Fecal *coliforms* and *Escherichia coli.*

For coagulation/flocculation using moringa seed powder, 250 mL of water from the cisterns was placed in five 250 mL Erlenmeyer flasks on a conventional shaking table. A concentration of 2.5 mL of the coagulant was then added to each one and the coagulation/flocculation time was 30 min. This procedure was carried out in triplicate.

After the coagulation/flocculation test, the agitator was turned off and the samples were left to rest for 45 min, so that the flocculated material could decant/sediment. The following parameters were then analyzed: electrical conductivity, pH, probable residue, $CaCO_3$, NaCl and total alkalinity in the Water and Soil laboratory of the IFPB Campus Sousa.

4.4 Variables analyzed

The moringa seeds used in the work were purchased from the IFPB - Campus Cajazeiras, Paraiba - Brazil.

The variables analyzed for the chemical composition of the seeds were: protein, lipids, ash and moisture.

The following parameters were also analyzed before and after water treatment: electrical conductivity, pH, probable residue, $CaCO_3$, NaCl and total alkalinity.

4.5 Statistical analysis

The results were subjected to analysis of variance (F test) and the means of the data were compared using the Tukey test at 5% probability, using the SISVAR statistical program (FERREIRA, 2000).

5. RESULTS AND DISCUSSION

5.1. Water treatment

Table 2 shows the results of the physical and chemical analysis of the parameters of Electrical Conductivity and pH before and after the treatment of the water from the cisterns.

Table 2. Results of physico-chemical analysis of EC and pH parameters before and after treatment of cistern water. Sousa - PB, 2016.

Sources	EC		pH	
	Before treatment	**After treatment**	**Before treatment**	**After treatment**
C1	0,15c	0,15c	6,23c	7,01c
C2	0,12d	0,12d	7.00bc	7,50b
C3	0,11d	0,11d	7,36b	7,53b
C4	0,35a	0,34a	8,60a	8,30a
C5	0,22b	0,21b	8.10ab	7,56b
DMS	0,01	0,01	1,31	0,29
CV (%)	2,86	2,07	6,23	1,39

For each attribute, means followed by equal letters do not differ by the Tukey test at 5% significance.

However, it can be seen that there was a significant difference between the average values for EC before the treatment of the cistern water, ranging from 0.35 dS m^{-1} to 0.11 dS m^{-1} , where C4 stood out with the highest value and C3 with the lowest value, with no significant difference between the values of C2 and C3.

For the EC parameter, after treating the water with the *Moringa oleifera* seed coagulant, it was observed that the values remained practically the same. However, these values are within the limit required by the WHO, which is 1 dS m^{-1} (BRASIL, 2011).

The pH of the water in the five cisterns before treatment ranged from low acidity (6.23) to high alkalinity (8.60).

All the analyses showed significant differences before and after treatment. However, after treatment with the coagulant obtained from moringa seed powder, the pH of the water ranged from neutral (7.01 $mg.L^{-1}$) to highly alkaline (8.30 $mg.L^{-1}$), with a decrease in these values after treatment. It can be seen that the use of the coagulant brought the pH of the water closer to neutrality, which is recommended for human consumption.

It can also be seen that the pH of the treated water was practically constant, as expected with the use of moringa as a coagulating agent (NDABIGENGESERE; NARASIAH, 1996). Santos (2007) did not obtain significant pH variations using moringa seeds in water. In practice, this implies a reduction in the cost of adding chemical products to correct pH when other substances are used in water treatment, such as aluminum sulphate.

WHO recommended values pH 7 to 8.5 for raw water and
6,5 8.5 for treated water because eye irritation and exacerbation of skin problems occur when it is higher than 11. As *Moringa oleifera* is a natural product, its application, especially in wastewater, is described as promising, as it can reduce the cost of chemical products used to adjust pH (BHATTI et al., 2007).

The decline in pH as a function of time and seed powder concentration was also verified by Borba (2001) and Oliveira et al.

(2011). Abaliwano et al. (2008), evaluating the benefits of Moringa, stated that M. oleifera does not significantly alter the pH and alkalinity of water after treatment (NDABIGENGESERE and NARASIAH, 1998; PISE and HALKUDE, 2012).

Table 3 shows the results of the physico-chemical analysis of the parameters Probable Residue and calcium carbonate ($CaCO_3$) before and after treatment of the water from the cisterns.

Table 3. Results of the physico-chemical analysis of the parameters Probable Residue and calcium carbonate ($CaCO_3$) before and after treatment of the water from the cisterns. Sousa - PB, 2016.

Sources	Residue Likely		CaCO3	
	Before treatment	**After treatment**	**Before treatment**	**After treatment**
C1	1073,33c	1067,66c	83,25c	85,00b
C2	860,00d	884,00d	74,40d	67,00c
C3	797,00d	850,33d	74,06d	66,00c
C4	2330,00a	2272,33a	162,6a	147,00a
C5	1490,00b	1494,00b	114,13	94,33b
DMS	96,13	44,93	7,69	10,60
CV (%)	2,60	1,21	2,68	4,09

For each attribute, means followed by equal letters do not differ by the Tukey test at 5% significance.

For Probable Residue, there was a significant difference between the average values before treatment, ranging from 2330.00 $mg.L^{-1}$ to 797.00 $mg.L^{-1}$, for C3 and C4, respectively. There was no significant difference between the average values of probable residue for cisterns C2 and C3.

After treatment, there was a decrease in probable residue values for treatments C1 and C4, while the opposite was observed in treatments C2, C3 and C5, where there was an increase in values. Legislation establishes a maximum permitted probable residue of 1000 mg.L^{-1} . However, it can be seen that treatments C1, C4 and C5 are outside this maximum permitted limit. Treatments C4 and C5 had the highest averages, which is probably because the water used to supply the cisterns was collected from a water reservoir by water tankers.

It can be seen that the $CaCO_3$ of the water before treatment ranged from 74.06 mg.L^{-1} to 162.61 mg.L^{-1} in cisterns C2 and C4, respectively.

After treating the water, it can be seen that the CaCO3 of treatments C2, C3, C4 and C5 decreased with the use of the natural coagulant extracted from the moringa seed powder, unlike treatment C1, which saw an increase from 83.25 mg.L^{-1} before treatment to 85.00 mg.L^{-1} after treatment.

Table 4 shows the results of the physico-chemical analysis of the chloride (NaCl) and total alkalinity parameters before and after treatment of the water from the cisterns.

Table 4: Results of physico-chemical analysis of chloride (NaCl) and total alkalinity parameters before and after treatment of cistern water. Sousa - PB, 2016.

Sources	NaCl		Alkalinity	
	Before treatment	**After treatment**	**Before treatment**	**After treatment**
C1	67,53c	149,90c	83,25c	85,00b
C2	53,53d	126,13d	74,40d	67,00c
C3	50,60e	113,20e	74,06d	66,00c

C4	164,40a	336,66ª	162,61a	147,00a
C5	103,86b	213,00b	114,13b	94,33b
DMS	2,91	8,34	7,69	10,60
CV (%)	1,18	1,57	2,68	4,09

For each attribute, means followed by equal letters do not differ by the Tukey test at 5% significance.

Analyzing the average NaCl values in the table, it can be seen that there was a range from 50.60 mg.L^{-1} to 164.40 mg.L^{-1} before water treatment, with treatment C4 standing out. After treatment, there was a significant difference in all the average values, which increased with the use of the natural coagulant extracted from moringa seed powder.

The total alkalinity of a water is given by the sum of the different forms of alkalinity, i.e. the concentration of hydroxides, carbonates and bicarbonates, expressed in terms of calcium carbonate. It can be said that alkalinity measures the water's ability to neutralize acids. Most natural waters have alkalinity values in the range of 30 to 500 mg.L^{-1} of $CaCO_3$, which is the unit used to express this parameter.

In the present study, the alkalinity of the water before treatment ranged from 74.06 mg.L^{-1} to 162.61 mg.L^{-1} , after treatment it was observed that treatments C2, C3, C4 and C5

decreased with the use of the natural coagulant extracted from moringa seed powder, unlike what happened with treatment C1, where an increase was observed from 83.25 mg.L^{-1} before treatment to 85.00 mg.L^{-1} after treatment. Waters with high alkalinity are unpleasant to the taste and the combination of high pH and excessive hardness can be harmful.

Ritter (2013) also observed a change in the water's chemical characteristics after treatment, as there was an increase in the alkalinity parameters in his study. However, even with this increase, the parameters remained within the values required by the Ministry of Health Ordinance No. 2914 of 12/12/2011, which sets out the procedures for controlling and monitoring the quality of water for human consumption and its potability standard.

Tables 4 and 5 show the results of the microbiological analysis of the water before and after treatment with the coagulant obtained from the moringa seed powder.

Table 5 - Fecal coliform analysis results before and after water treatment with moringa seeds. Sousa-PB, 2016.

	Fecal coliforms	
Treatments	**Before Treatment**	**After Treatment**
C1	35° C> 1.1 X 103	35° C> 1.1 x 10^3
C2	35° C> 1.1 x 10^3	35° C> 1.1 x 10^3
C3	35° C> 1.1 x 10^3	35° C> 1.1 x 10^3
C4	35° C> 1.1 x 10^3	35° C> 1.1 x 10^3
C5	35° C> 1.1 x 10^3	35° C> 1.1 x 10^3

The result for fecal coliforms remained the same after treating the water with the moringa coagulant. Drinking water should be free of pathogenic microorganisms and bacteria that indicate fecal contamination.

According to Ministry of Health regulation 2914, the presence of coliforms in water indicates contamination, with the potential risk of the presence of pathogenic organisms, and since they are more resistant in water than pathogenic bacteria of intestinal origin (BRASIL, 2011).

Table 6. Results of analysis for *Escherichia coli* before and after treatment of water with moringa seeds. Sousa-PB, 2016.

Escherichia coli		
Treatments	**Before Treatment**	**After Treatment**
C1	Presenga	Presenga
C2	Absence	Absence
C3	Absence	Absence
C4	Absence	Absence
C5	Presenga	Presenga

Of the five treatments, two (C1 and C5) showed the presence of *Escherichia coli* both before and after treating the water with the coagulant obtained from the moringa seed powder (Table 6). The water in treatment C1 was rainwater, so it wasn't expected to contain microbiological contamination because rainwater is understood to be free of contaminants. This contamination could probably have been caused by animals coming into contact with the water, the pipes used to collect the water could have been dirty with animal waste, or the container used to collect the water could have been contaminated.

Treatment C5 was water from a cistern supplied by a water tanker, which could probably be contaminated. Cleaning this type of water transportation is essential to avoid contaminating the water and, consequently, those who use it. The other treatments showed no *Escherichia coli*. Drinking water is considered to be water with no *E. coli* or fecal coliforms per 100 mL of sample (BRASIL, 2011).

According to Rorato (2013), it is possible to verify that *Moringa oleifera* Lam is a natural polymer that can be used as a coagulation aid in the experimental conditions that were analyzed, since good results were obtained in the removal of total coliforms and *E. coli*, in the combination of Poly Aluminum Chloride (PAC) with moringa, since both became absent after water treatment.

6. CONCLUSIONS

The values found for the chemical composition of moringa seeds were: proteins 34.22 %, lipids
39.84 %, ash 3.35 % and moisture 5.02 %. Generally, protein is the compound found in greatest quantity in *M. oleifera* seeds, and in this study the compound found in greatest quantity was lipid, since moringa seeds have a high oil content, with values varying according to the variety of plant, climatic conditions, ripening time, harvesting season and extraction method used.

The dosage of coagulant used was not enough to alter the presence of fecal *coliforms* and *Escherichia coli.*

Moringa oleifera Lam. seeds can be used as an effective natural coagulant in simplified water treatment.

The use of coagulants extracted from moringa seeds is feasible for simplified water treatment, being a low-cost procedure and not involving the use of chemical additives normally present in conventional coagulants.

Considering that moringa can be produced locally, we recommend the production of this plant and its use as a coagulant in the treatment of water from cisterns and other water sources.

The use of coagulants of plant origin for water treatment, when possible, is of great value in ecological and environmental terms. Moringa therefore stands out as a promising natural coagulant.

Further tests with different amounts of the moringa seed coagulant and different coagulation/flocculation times are recommended.

The use of moringa as a natural coagulant in water treatment can help reduce the number of waterborne diseases, improve hygiene habits and quality of life and contribute to the sustainable development of the region.

7. REFERENCES

ABALIWANO, J. K.; GHEBREMICHAEL, K. A.; AMY, G. L. Application of the purified *Moringa oleifera* coagulant for surface water treatment. WaterMill Working Paper Series, n. 5, p. 1-19, 2008.

AMAGLOH, F. K., BENANG, A. Effectiveness of *Moringa oleifera* seed as coagulant for water purification. African Journal of Agriculture Research, v.4, n.1, p.119-123, 2009.

AYERZA, R. Seed and oil yields of Moringa *oleifera* variety Periyakalum-1 introduced for oil production in four ecosystems of South America. Industrial Crops and Products. v. 36, p. 70-73. 2012.

BARRADAS, J. L. D. Tanino - Uma solugao ecologicamente correta: agente floculante biodegradavel de origem vegetal no tratamento de água. Novo Hamburgo: Technical Publication, 2004.

BHATTI, H., MUMTAZ, B., MUMTAZ, B. Removal of Zn (II) ions from aqueous solution using *Moringa oleifera* Lam. (horseradish tree) biomass. Process Biochemistry, Vol. 42, pp. 547-553, 2007.

BHOLE, A. G. Relative evaluation of a few natural coagulants. Aqua, vol. 44, n° 6, pp. 284-290, 1995.

BINA, B; MEHDINEJAD, M.H; NIKAEEN, M; MOVAHEDIAN, H. Attar. Effectiveness of chitosan as natural coagulant aid in treating turbid waters. Department of Environmental Health Engineering, School of public health, Isfahan University of Medical Sciences, Isfahan, Iran, Vol 6, N° 4, p 247-252, 2009.

BONGIOVANI, M. C.; MORAES, L. C. K.; BERGAMASCO, R.; LOURENQO, B. S. S.; TAVARES, C. R. G. The benefits of using natural coagulants to obtain drinking water. Departamento de engenharia quimica, Universidade Estadual de Maringa, v 32, n 2, p 167-170, 2010.

BORBA, L. R. Viabilidade do uso da *Moringa oleifera* Lam no Tratamento Simplificado de Agua para Pequenas Comunidades. Federal University of Paraiba, master's thesis, 92p, 2001.

BOTERO, W. G. Characterization of sludge generated in water treatment plants: prospects for agricultural application. Chem. Nova, Vol. 32, No. 8, 2018-2022, 2009.

BRAZIL. Ministry of Health. Ordinance no. 2914 of December 12, 2011. Provides for procedures to control and monitor the quality of water for human consumption and its standard of potability. Diario Oficial da Uniao, Brasilia, 2011.

BRAZIL. MINISTRY OF HEALTH. Surveillance and control of the quality of water for human consumption. Brasilia: Secretaria em Vigilancia em Saude, 212p., 2006.

CARDOSO K. C. Estudo do Processo de Coagulapao/Floculapao

por meio da *Moringa oleifera* Lam para Obtenpao de Agua Potavel. Master's thesis, Parana-Brazil, 123 f. 2007.

CARRIJO, K. C. Study of the use of biodegradable natural flocculants in water clarification. Final thesis: Chemical Engineering, Federal University of Uberlandia - UFU, Uberlandia, Minas Gerais, Brazil 2002.

CARVALHO, M. J. H. Use of natural coagulants in the process of obtaining drinking water. Master's dissertation. State University of Maringa, p. 23, 2008.

CLAYTON, B. E. Report of the Lower moor Incident Advisory Group. J. Ind. Med., London, v. 40, n. 3, p. 301-304, 1989.

DAHOT, M. U. Vitamin contents of the flowers and seeds of *Moringa oleifera* L. Journal of Biochemistry, v.21, n.1-2, p.21-24, 1998.

DALLA ROSA, K. R. Morhga a: a perfect tree for home gardens. Hawaii: NFTA, 2p, 1993.

DI BERNARDO, L. **Metodos e Tecnicas de Tratamento de Agua**. Abes, Rio de Janeiro, 1993.

DI BERNARDO, L., DANTAS, A. D. B. Metodos e tecnicas de tratamento de agua. 2.ed, Sao Carlos: Rima Editora, v.2, p.1584, 2005.

DUKE, J. A. The Quest of Tolerant Germplasm. In: YOUNG, G. (Ed.) Crop tolerance to subtropical land conditions. Madison. American Society Agronomial Special Symposium, v.32, p.1-16, 1978.

FERREIRA, D. F. Statistical analysis using SISVAR (System for Variance Analysis) for Windows version 4.0. In: ANNUAL MEETING OF THE BRAZILIAN REGION OF THE INTERNATIONAL SOCIETY OF BIOMETRY, 45, 2000, Sao Carlos. Proceedings... Sao Carlos: UFSCar, p. 255-258, 2000.

FUGLIE, L. J. The Tree Miracle: Moringa oleifera: Natural Nutrition for the Tropics. Church World Service, Dakar. Church World Service, Dakar. 68PP., Revised in 2001 and published as The Miracle Tree: Moringa's many attributes, 172 pp.

FUNASA - MINISTRY OF HEALTH/NATIONAL HEALTH FOUNDATION. Toxic Cyanobacteria in Drinking Water. Impacts on Public Health and Removal Processes in Drinking Water. FUNASA/MS, Brasilia, Brazil, 56p, 2003.

GALLAO, M. I., LEANDRO, F. D., BRITO, E. S. Chemical and structural evaluation of Moringa seed. Revista Ciencias Agrarias, Fortaleza, v.37, n.1, p.106- 109, 2006.

GASSEN, H. G., GASSENSCHIMIDT, U., JANY, K. D., TAUSCHER, B., WOLF, S. Modem Methods in protein and nucleic acid analysis. Biological Chemistry Hoppe-Seyler, p.757-769, 1990.

GERDES, G.; **Use of the Seeds of the *Moringa oleifera* Tree for the Treatment of Turbid Waters**. ESPLAR, Fortaleza, 1996.

GHEBREMICHAEL, K. A., GUNARATNA, K. R., HENRIKSSON, H., BRUMER H., DALHAMMAR, G. A simple purification and activity assay of the coagulant protein from *Moringa oleifera* seed. Wat. Res., v.39, p 2338-2344, 2005.

GRAHAM, N., GANG, F., FOWLER, G., WATTS, M. Characterization and coagulation performance of a tanninbased cationic polymer: a preliminary assessment. Colloids and Surface A: Physicochemical and Engineering Aspects, v. 327, n. 1-3, p. 9-16, 2008.

HONG, H. C., LIANG, Y., HAN, B. P., MAZUMDER, A., WONG, M. H., JAHN, S. A. A. Monitored water coagulation with Moringa seeds in village households. GTZ: Gate. Eschborn, n. 1, p. 40-41, 1989.

BRAZILIAN INSTITUTE OF GEOGRAPHY AND STATISTICS - IBGE. National basic sanitation survey 2008. Rio de Janeiro: MPOG, MC, 219p., 2010.

JAHN, S. A. A. Using Moringa seeds as coagulant in developing countries. Journal of the American Water Works Association, vol. 6, pp. 43-50, 1988.

KARADI, R. V., GADGE, N. B., ALAGAWADI, K. R., SAVADI, R. V. Effect of *Moringa oleifera* Lam. root-wood on ethylene glycol induced urolithiasis in rats. Journal of Ethno pharmacology, v. 105, p. 306-311, 2006.

KATAYON, S.; NOOR, M. J. M. M.; ASMA, M.; ABDUL, G. L. A., THAMER, A. M.; AZNI, I.; AHMAD, J.; KHOR, B. C.; SULEYMAN, A. M. Effects of storage conditions of *Moringa oleifera* seeds on its performance in coagulation. Bioresource Technology, v. 97, p. 14551460, 2006.

LIBANIO, M. **Fundamentos de qualidade e tratamento de agua**. 2 Ed., Campinas: Editora Atomo, 444p., 2008.

LILLIEHOOK, H. Use of Sand Filtration of River Water Flocculated with Moringa oleifera. Master's Thesis, 27p. Department of Civil and Environmental Engineering, Division of Sanitary Engineering. Lulea University of Technology, Lulea, 2005.

LO MONACO, P. A. V.; MATOS, A. T.; RIBEIRO, I. C. A.; NASCIMENTO, F. S.; SARMENTO, A. P. Utilization of Moringa seed extract as a coagulating agent in the treatment of water supply and wastewater. Ambi-Agua, Taubate, v. 5, n. 3, p. 222-231, 2010. doi:10.4136/ambi-agua.164. Available

in: <
http://www.amigosdanatureza.org.br/publicacoes/index.php/forum_ambiental/article/viewFile/557/582>. Accessed on: February 28, 2016.

LORENZI, H., MATOS, F. J. A. Plantas Medicinais no Brasil: Nativas e Exoticas Cultivadas. Nova Odessa: Instituto Plantarum, p. 347, 2002.

LORGE, E., FELCH, C., PATTERSON, M., KUMFER, B. Redesigned filter for produced water treatment.Surface Systems and Solutions. 2011. Siemens Industry Inc. Available at: <http://www.water.siemens.com/SiteCollectionDocuments/Product_Lines/Monosep/Brochures/EP_July_2011_Redesigned_Filter_PW_Trt.pdf>. Accessed on September 25, 2016.

MANGRICH, A. S., DOUMER, M. E., MALLMANN, A. S., WOLF, C. R. Green Chemistry in Water Treatment: Use of Coagulant Derived from Acacia mearnsii Tannin. Revista Virtual de Quimica, Curitiba-PR, 2013.

MORAES, L. C. K. Study of Coagulation-Ultrafiltration with the Biopolymer Chitosan for the Production of Potable Water. Master's dissertation, Department of Chemical Engineering, State University of Maringa - UEM, Maringa, Parana, Brazil 2004.

MUYIBI, S. A., EVISON, L. M. *Moringa oleifera* seeds for softening hardwater. Water Research, vol. 29, n° 4, pp. 1099-1104, 1995.

NDABIGENGESERE A., NARASIAH, S. K. Influence of operating parameters on turbidity removal by coagulation with *Moringa oleifera* seeds. Environmental Technology, v.17, p.1103-1112, 1996.

NDABIGENGESERE A., NARASIAH, S. K; TALBOT B. G. Active

agents and mechanism of coagulation of turbid waters using *Moringa oleifera.* Water Research, v.29, n.2, p.703-710, 1995.

NDABIGENGESERE, A. NARASIAH, K. S. Quality of water treatedby coagulation using *Moringa oleifera* seeds. Water Research, v.32, n.3, p.781-791, 1998.

OKUDA, T., BAES, A. U., NISHIJIMA, W., OKADA, M. Improvement of extraction method of coagulation active components from *Moringa oleifera* seed. Water Research, v.33, p.3373-3378, 1999.

OLIVEIRA, I. C.; TEIXEIRA, E. M. B.; GONQALVES, C. A. A.; PEREIRA, L. A. Centesimal evaluation of Moringa oleifera Lam seed. II Seminario Iniciagao Cientifica - IFTM, Campus Uberaba, MG. October 20, 2009.

PAN AMERICAN HEALTH ORGANIZATION. Water and Health. Available at: <http://www.opas.org.br/ambiente/UploadArq/agua.pdf>. Accessed on: September 18, 2016.

OZACAR, M., SENGIL, I. A. Adsorpiton of acid dyes from aqueous solutions by calcined alunite and granular activated carbon. Adsorptions 8:301-308, 2002.

PARENTE, E. J. S. **Biodiesel: A technological adventure in a funny country**. Fortaleza: Unigrafica, 2003.

PEREIRA, D. F. Potential *of Moringa oleifera* Lam in the production of biodiesel and in the treatment of water produced in oil extraction. Sao Cristovao, 2011.

PEREIRA, D. F., VASCONCELOS, V. M., VIEIRA, A. C., ARANDA, D. A. G., da SILVA, G. F. Production of biodiesel from the hydro esterification of moringa oil. II National Moringa Meeting, November 3 to 5, 2010, Aracaju - SE.

PISE, C. P.; HALKUDE, D. S. A. Blend of natural and chemical coagulant for removal of turbidity in water. International Journal of Civil Engineering and Technology, v. 3, n. 2, p. 188-197, 2012.

RANGEL, M. S. *Moringa oleifera:* A natural water purifier and food supplement for northeastern Brazil. Aracaju-SE, 2004. Available at: <http: // www.jardimdeflores.com.br/floresefolhas/A10moringa.htm>. Accessed on: October 11, 2016.

RARNACHANDRAN, C., TER, K. V., GOPALAKRISHMAN, P. K. Drumstick *(Moringa oleifera):* A multipurpose Indian vegetable. Economic Botany. v.34, p.276-283, 1980.

RASHID, U., ANWAR, F., ASHRAF, M., SALEEM, M., YUSUP, S. Application of response surface methodology for optimizing transesterification of *Moringa oleifera* oil: Biodiesel production. Energy Conversion and Management 52, 3034-3042, 2011.

RITTER, C. M. Study of the use of natural polymers Abelmoschus esculentus (l.) Moench (Malvaceae) and *Moringa oleifera* Lam (Moringaceae) in the treatment of water supply. FEDERAL TECHNOLOGICAL UNIVERSITY OF PARANA, Campo Mourao - Parana, 2013. Available at: < http://repositorio.roca.utfpr.edu.br/jspui/bitstream/1/1599/1/CM_COEAM_2013_1_08.pdf> Accessed on: March 02, 2016.

RONDEAU, V.; COMMENGES, D.; JACQMINGADDA, H.; DARTIGUES, JF. Relation between aluminum concentrations in drinking water and Alzheimer's disease: an 8-year follow-up study. American Journal of Epidemiology, v. 152, n. 1, p. 59- 66, 2000.

RORATO, W. R. Use of *Moringa oleifera* Lam. as an aid in the Coagulation/Flocculation/Filtration process for the treatment of water supply. FEDERAL TECHNOLOGICAL UNIVERSITY OF PARANA, Campo Mourao - Parana, 2013. Available at: <http://repositorio. roca.utfpr.edu.br:8080/jspui/bitstream/1 /1626/1/CM _COEAM_2013_1_27.pdf>. Accessed on: March 2, 2016.

SANTANA, C. R. Treatment of produced water through the flotation process using *Moringa oleifera* Lam as a natural coagulant. Sao Cristovao. UFS/PEQ. 2009. 155 p.
(Master's dissertation).

SANTOS, O. R. "Use of *Moringa oleifera* seeds for the treatment of textile effluents." In: 24th CONGRESS
BRAZILIAN ENGINEERING, 2007.

SILVA, F. J. A., MATOS, J. E. X. On *Moringa oleifera* dispersions for water treatment. Revista Tecnologia, v.29, n.2, p.157-163, 2008.

SILVA, T. S. S. Study of physico-chemical treatability with the use of vegetable tannins in water supply and sewage. Rio de Janeiro: Oswaldo Cruz Foundation; National School of Public Health, 1999.

VASCONCELOS, V. M., SILVA, P. C. G., VIEIRA, A. C., SILVA, G. F.;Production of Biodiesel 'in situ' from *Moringa oleifera* Lam. XXXIV Brazilian Congress of Particulate Systems. Campinas - SP, 2009.

ZHANG, Z., HALL, A., PERFECT, E., GURR, S. X. Differential expression of two Blumeria graminis chitin synthase genes. Mol. Plant pathol. 1, 125-138, 2002.

Index

Printed by Books on Demand GmbH, Norderstedt / Germany